ATOMIC AND MOLECULAR
ELECTRONIC
CONFIGURATION
REVISITED

Harold J. Teague

CONTENTS

INTRODUCTION

What is found in the pages of AMEC is an attempt to show that Chemistry is a journey, specifically involving the areas surrounding the Noble Elements of the Periodic Table. The first sections concentrate mostly on the area between the Noble Elements He and Ne, the Row Two Elements, and related diatomic structures.

Make no mistake about it, chemistry *is* 'the areas' before and between the noble elements and can be seen as the progression from one noble element to another noble element. For instance, see Chapter 3: *Selections From Atomic and Molecular Electronic Configuration 1987*[©] page (82) 57 for a plot of Number of Chemical Bonds versus Number of Electrons for row two species. It will be seen that N_2, near the middle, has the maximum number (3.0) of bonds. This also means that N_2 does not contain any 'non-bonded' orbitals from the original N atoms. Fittingly, although problems related to the O_2 molecular structure was, and is, the stimulus for this section, it will begin with NO[¹] which has the same electronic configuration as N_2 (10 Valence Electrons).

The oxygen, O_2 molecular structure (12 Valence Electrons) has been one of the more difficult problems to solve in all of chemistry. However, it is not only the O_2 structure that presents problems but also its Row Two neighboring elemental diatomic species of the Periodic Table, specifically the NO structure (11 Valence Electrons) and the uncharacterized (?) OF molecular structure (13 Valence Electrons). These three diatomic species make up a *Lone P Orbital Pi (π) Bond* companion set just as N, O, and F are the only three elements capable of Hydrogen Bonding. The basis for both Hydrogen Bonding and the Lone P Orbital Pi (π) Bond Systems is the same: These three elements, N, O, and F, solely meet the necessary requirements of being both *small* and *electronegative*.

In the early 1980's two model systems for arriving at molecular structure by General Chemistry texts were the LCAO-MO (Linear Combination of Atomic Orbitals-Molecular Orbital) and MOO (Maximum Orbital Overlap) Model systems. The MOO model led to the principle that p orbitals tended to overlap end-to-end. As is seen in the 1987 version and the current AMEC version this concept presented a number of problems, especially with the B_2 and C_2 (Ground State) structures.

A number of questions of the type above led this writer to self-publish in 1987 *Atomic and Molecular Electronic Configuration* (and present many talks and workshops) related to chemical structure. Two systems that created difficulty then, and to this day, are the NO and O_2 structures. The treatment in 1987 can be seen in Chapter 3 (*Selections From Atomic and Molecular Electronic Configuration 1987*[©]). In this updated (2014) version of AMEC, the focus is on Row Two elements (Periodic Table) and molecules containing these elements (and hydrogen). In the study of diatomic species in Chapter 1, Section V it will be seen that the trend in bonding (involving s and p orbitals) reaches a maximum with N_2 (: N $\equiv$ N :) with a Bond Order (BO) of 3.0. Thereafter, the bond order decreases as p orbitals are re-formed (and hybridization occurs). Finally, it will be seen that if an electron is added to F_2 the atomic species F^0 and F^- (isoelectronic with Ne) forms.

CHAPTER 1

PART A. THE LONE P ORBITAL PI BOND SYSTEM:
NO, O_2, and OF STRUCTURES

PART B. ROW TWO DIATOMIC SEQUENCE: $LiBe > F_2$

ANALYZING MOLECULAR STRUCTURE

For a proposed structure to be 100% correct for a specific substance, it must adhere to the established facts known for that substance. Two items that usually are known and must be accounted for are Bond Order (BO) and Paramagnetic Number (P)

1. BOND ORDER (BO). This relates to the bond strength that holds the atoms together. The number of bonding electrons, sigma (σ) and pi (π) summed and divided by two gives the bond order. A single orbital, bonding or atomic, may possess an electron pair or a single electron. For example, N_2 has a BO = 3.0, comprising two pi bonds and one sigma bond.

2. PARAMAGNETISM (P). An atom or molecule may possess one or more unpaired electrons (however, a single orbital may possess only a single unpaired electron). For example, the O_2 molecule possesses two unpaired electrons and is, therefore, paramagnetic; it reacts to magnetic fields.

THE LONE P ORBITAL PI (π) BOND SYSTEM

THE LONE P ORBITAL PI BOND SYSTEM. The NO, O_2, and OF systems involve non-traditional pi (π) bonds labeled the *Lone P Orbital Pi (π) Bond* and defined as follows: one p orbital of a single atom forms the pi bond. As will be observed, this pi bond is very different from the normal π bond, which involves side-to-side overlap of aligned p orbitals, each p orbital coming, respectively, from the two atoms of the pi bonding system. The *Lone P Orbital Pi Bond*, however is limited to bonding systems containing only N, O, and F. In the *Lone P Orbital Pi Bond* system, as noted above, an unpaired p orbital electron of one atom forms a *lone p orbital pi bond.* It does so by being attracted into the pi bonding region by the strong electronegative character of the second (N, O or F) atom. However, the electronegative character alone is not sufficient as the atom also must be small in size. The *Lone P Orbital Pi Bond* system is necessary as there is no p orbital of the second atom *available* for pi bonding. The total number of 'orbital' bonds for N, O and F is four, which would be violated with a traditional pi orbital. Again, it is stressed that both small size and high electronegative character are necessary, which are satisfied only by N, O, and F! It should be noted that these are the same conditions (and elements) involved in Hydrogen Bonding.

THE NO ELECTRONIC STRUCTURE

I. MRAE METHOD FOR THE NO ELECTRONIC STRUCTURE. The following data is known for NO: BO (bond order) = 2.5; P (paramagnetic) = 1. In Sections III and IV, MRAE-MO Energy-Level Diagrams/Template the process is to build on a previous structure. Since the N_2 molecule is isoelectronic with NO^+ and N_2 has a well established structure, it seems fitting to begin with the NO^+ species (one σ bond and two traditional π bonds). In addition, each atom (of NO^+) has all four of its outer (valence) shell orbitals ($2s^2 2p^1 2p^1 2p^1$) modified due to molecular bonding. The following sequence occurs when an electron (MRAE) is added to NO^+. (For clarity the overall process will be separated into three steps.)

    a. Disruption of a Pi bond. When an electron (MRAE), which has to be anti-bonding, adds to NO^+ it causes disruption of one of the traditional pi bonds. This occurs by the MRAE *pairing* with one of the pi electrons, which then moves into a p orbital of the O atom. This leaves a single electron in the pi bond, which now is associated with *only* a p orbital of the N atom; no longer is it a traditional pi bond. For convenience, however, this electron, for now, will be placed on the N atom. This intermediate form is possibly an excited state form (sometimes offered as the Lewis structure for NO). (When convenient, in the structures that follow only one pi orbital component will be shown.)

NO From NO⁺ (or N₂)

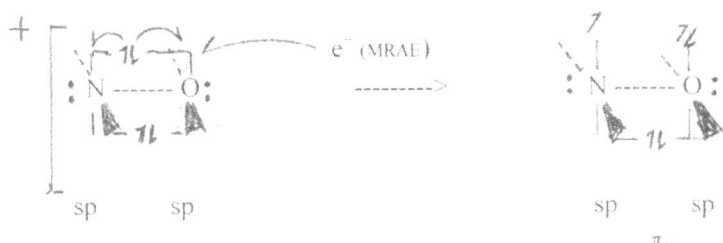

b. sp² Hybridization. At this point the species has only one traditional double bond. Therefore, the remaining adjacent atomic p orbitals are aligned with a total of three electrons. Since the O (and N) atom is small and electronegative this electron repulsion induces the O atom to become sp² hybridized. (See Chapter Two and AMEC, 1987 for treatment of hybridization.)

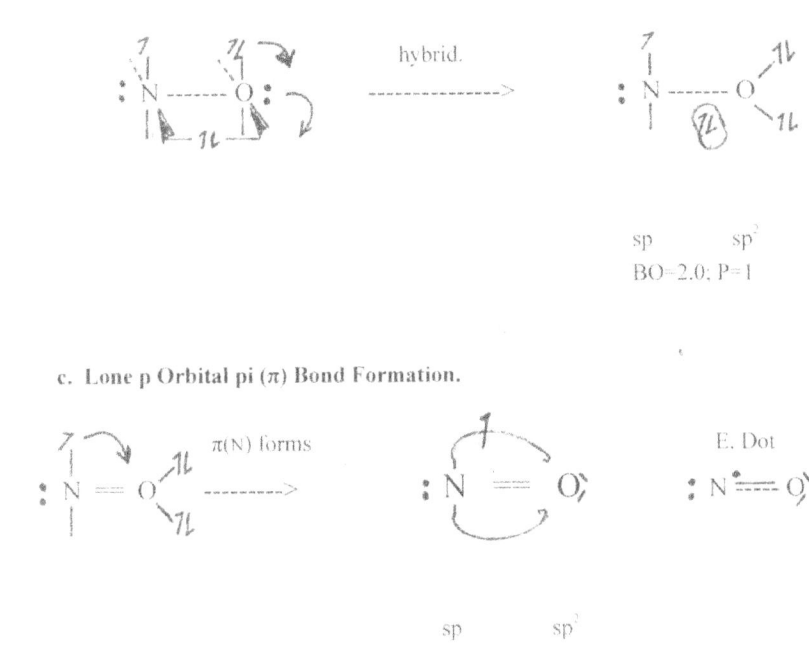

c. Lone p Orbital pi (π) Bond Formation.

It is emphasized that the N p orbital forms a pi (π) orbital, not by overlapping with another p orbital but by being attracted into the realm of the oxygen atom. *No orbital of the oxygen atom is used, nor available, for the pi bond.* The small size and strong electronegative character of the oxygen atom pulls the N p orbital electron into the pi bonding region, labeled $\pi(N)$. Again, it is stressed that no orbital of the oxygen atom is available for bonding as oxygen is limited to four orbitals.

THE LCAO (Linear Combination of Atomic Orbitals) MOLCEULAR ORBITAL METHOD. In the LCAO method the two atoms (N: $2s^2 2p^1 2p^1 2p^1$, etc) are brought together along the x-axis with the x-axis p orbital and the 2s orbital merging. This in effect produces sp hybridization as the other two p orbital are not altered, at this point. Symmetric about the x-axis are the sigma (σ) regions comprising a bonding region between the two nuclear cores and two equal energy regions outside the bonding region. In AMEC it is proposed that these regions are non-bonded regions, not anti-bonding regions. Next, depending upon the number of valence electrons, the p orbital may overlap side-to-side to form traditional pi (π) orbital bonds. (See Sections III and V for more in depth treatment of these concepts.)

II.   LCAO METHOD for NO. Its structure can be constructed from N and O as follows:

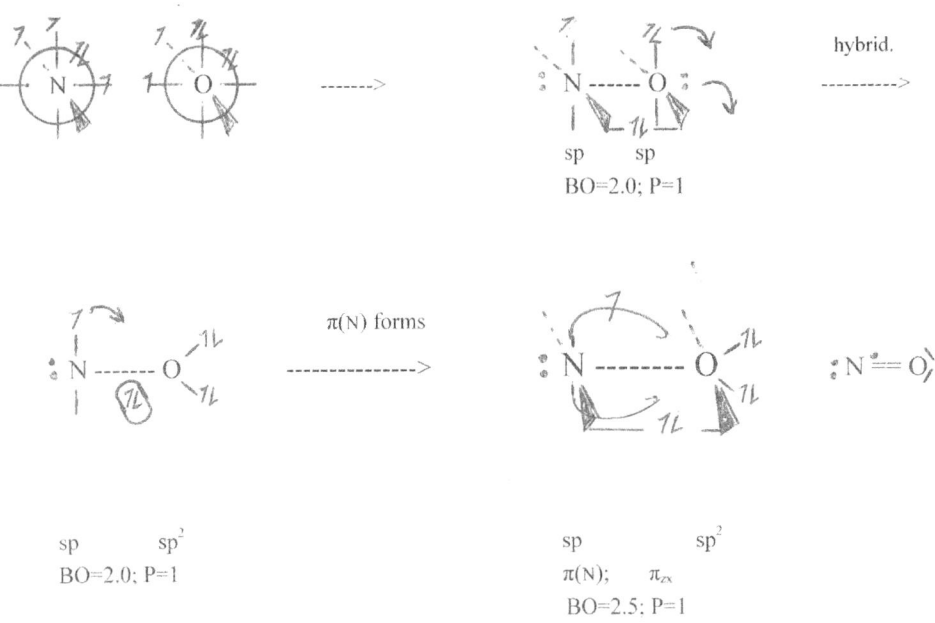

OXYGEN, O₂ STRUCTURE

As noted above, the oxygen, O_2 molecular structure (12 Valence Electrons) has been one of the more difficult problems to solve in all of chemistry. As already seen, it is not only the O_2 structure that presents problems but also the NO structure (11 Valence Electrons) as well as the uncharacterized (?) OF molecular structure (13 Valence Electrons). It is worth repeating: These three diatomic species make up a companion set just as N, O, and F are the only three elements capable of Hydrogen Bonding. The basis for both Hydrogen Bonding and the *Lone P Orbital Pi (π) Bond* is the same: These three elements are *small* and *electronegative!*

As discussed above for a proposed structure to be 100% correct for a specific substance, it must adhere to the established facts known for that substance. Let us review some of the information for O_2.

1. BOND ORDER (BO). The bond order is 2.0.

2. PARAMAGNETISM (P). There are two unpaired electrons (P = 2). That is, O_2 reacts to a magnetic field. One popular demonstration of paramagnetic character is to pour liquid oxygen between the pole of a strong magnetic. Strands of liquid oxygen are seen forming between the magnet's poles. It is possible the oxygen molecules are undergoing polymerization as follows:

3. HYBRIDIZATION. The hybridization of each oxygen atom is sp². In many cases it is difficult to know for certain the hybridization state of a species with only one atom bonded to it. However, the oxygen molecule binds to the Fe^{+2} heme complex of the hemoglobin molecule, whose structure has been determined through X-Ray crystallography. The oxygen loosely binds at an angle of about 120^0 and the structure does not appear to further hybridize due to the binding to the iron. This is strong evidence that the oxygen atom binding to the heme is sp² hybridized.

(Note: As the analysis of Row Two diatomic species progresses from left to right across the Periodic Table a trend in hybridization occurs: sp > sp² > sp³. The sp³ hybridization proposed for F atoms of F_2 is based on the small, electronegative character of the F atom. Not only are H-Bonding and Lone p Orbital Pi Bond systems argued to be based on the small, electronegative character of N, O, and F atoms , but likewise is hybridization.)

The importance of hybridization is also seen as both toxic CO and CN^-, which are isoelectronic with N_2 (sp hybridized) bind very strongly (about 200 times more strongly than does O_2) with the heme Fe^{+2} in a straight-on orientation, not at an angle. (Again, this is supported by X-Ray crystallography.)

Oxygen Bound To Heme (Hemoglobin)

4. OXIDATION. The O_2 does not appear to bridge across adjacent C atoms in a C --- C double bond. Instead, it often bridges three atoms (as it apparently does in prostaglandin synthesis), or it forms a peroxide species.

PAST/CURRENT MODELS FOR O_2 AND RELATED PROBLEMS. An overview of some current general chemistry texts shows that most authors do not attempt Lewis or Electron-Dot Structure(s) for O_2 and just skip over them. Instead, the Molecular Orbital (M O) model is offered. However, many authors continue to give the Lewis Structure for NO, which has the same problem as does O_2 (see Zumdahl, 5th ed.)

Two options for the Lewis structure for O_2 have been used. Let us analyze them and point out difficulties with them.

Option 1.
$$\ddot{O} --- \ddot{O}$$
BO=1.0; P=2

Option 2.
$$\ddot{O} \doteq \ddot{O}$$
BO=2.0; P=2

Problem(s) With Option 1. This Lewis Structure actually represents the precursor to the correct O_2 structure. Its problem is, obviously, that it has a bond order (BO) of 1.0, not 2.0 which is well established. It does, however, indicate that it is paramagnetic (P=2).

Problem(s) With Option II. The bond order (BO = 2.0) is correct but there is a problem with the two (paramagnetic) unpaired electrons. Unless some explanation to the contrary is given, it has to be assumed that each unpaired electron is in a traditional pi orbital. That means that each oxygen must have five orbitals: one sigma bonding "sp" component, two orbitals for each non-bonding electron pair, and two p orbitals for the two paramagnetic electrons. However, oxygen as a row two element can only have four valence shell (outer shell) orbitals, not five. The O atom has one 2s and three 2p orbitals and thus is limited also to four molecular orbitals.

AN OXYGEN, O_2 STRUCTURE CONSISTENT WITH THE DATA.

I.   MRAE Method For O_2, From O_2^+ (See NO above).

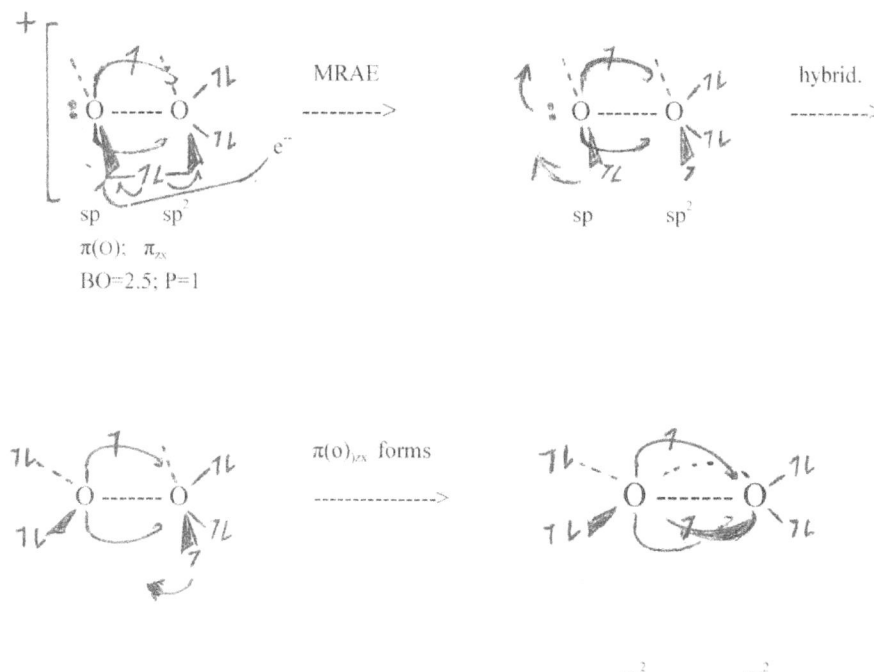

II. LCAO METHOD FOR O_2.

hybrid.

-------------->

sp   sp

$\pi(o)$'s form

----------->

E. Dot

sp^2 sp^2

$\pi(o)_{jyx}$ $\pi(o)_{zx}$

BO=2.0; P=2

See the following sections also:

III.. MECHANISTIC SEQUENCE FOR: $NO^+ > NO > O_2 > OF > F_2 > F^0 / F^-$.

IV. PHOTOGRAPHS OF NO AND O_2 MODELS.

V.. MRAE-MO Energy-Level Diagrams.

VI. MRAE-MO Energy-Level Diagrams Template.

COMPARISON OF O_2 AND $H_2C== CH_2$ STRUCTURES

O_2 and ethene, H_2CCH_2 have 12 valence electrons and, therefore, are isoelectronic. However, these two molecules are very different. Ethene forms a traditional double bond (two atoms, each contributing a p orbital, with one electron in each p orbital, overlapping side-to-side to form a pi orbital/bond). As seen above, each oxygen atom of the oxygen molecule has the two p orbitals at 90^0 to each other. Furthermore, each p orbital electron forms a pi orbital, not by overlapping with another p orbital, but by being 'attracted' into the realm of the other oxygen atom. *No orbital of the second oxygen atom is used.* The strong electronegative character of each oxygen atom pulls each p orbital electron into the respective pi bonding region! This accounts both for the Bond Order of 2.0 and for the oxygen molecule being paramagnetic (P=2), while ethene is diamagnetic (P=0).

THE OF ELECTRONIC STRUCTURE

The OF molecule has 13 valence electrons and, as with NO, it has an one p orbital pi system. The O atom is sp^2 hybridized and the F is sp^3 hybridized. The OF molecule has a BO of 1.5 and P=1.

I.   MRAE METHOD FOR OF FROM OF$^+$ (or O$_2$).

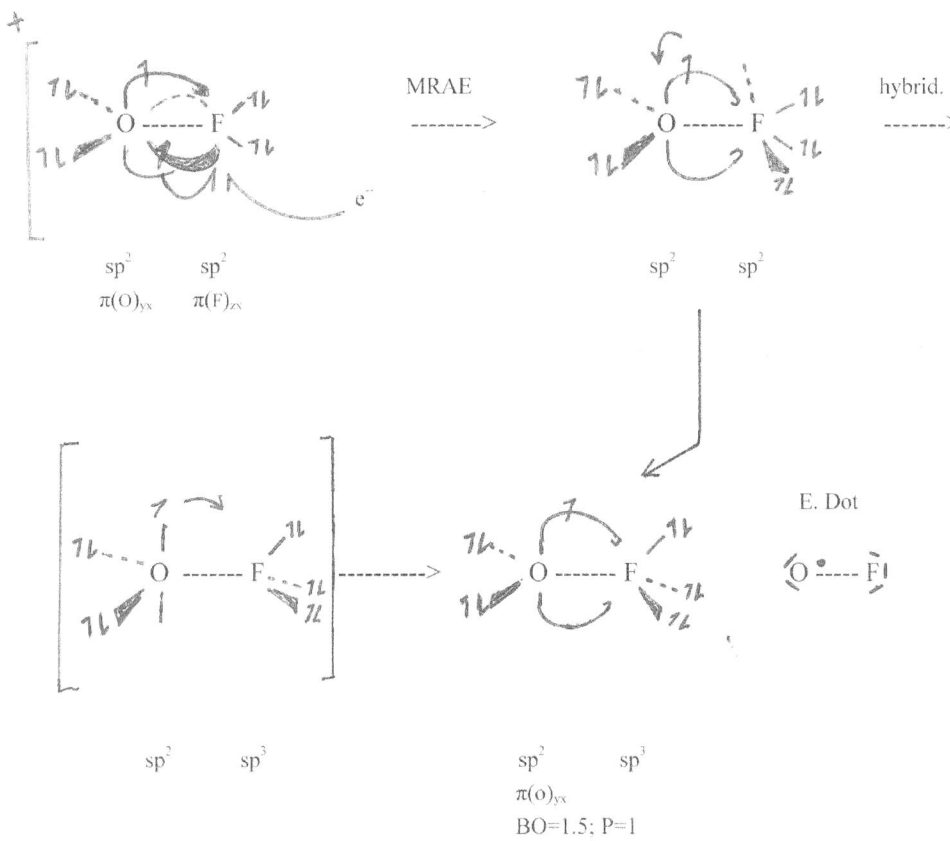

II.   LCAO METHOD FOR OF.

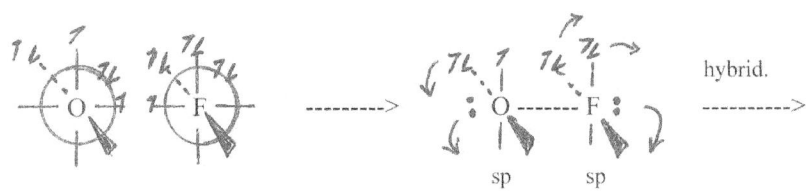

hybrid.
- - - - - - - >

sp   sp

π(o) forms
- - - - - - - >

sp²   sp³

sp²   sp³
π(o)ᵧₓ
BO = 1.5; P=1

See the following sections also:

III. MRAE MECHANISTIC SEQUENCE FOR:

NO^+ (or N_2) > NO (or O_2^+) > O_2 (or NO^- or OF^+) > OF (or O_2^-) > F_2 >

NO^+ (or N_2)	NO	NO (or O_2^+)
$+ \left[:N \equiv O: \right]$ e^-	$\longrightarrow$ $:N \equiv O:$	$\longrightarrow$ $:N \equiv O$
sp sp	sp sp	sp sp^2
		$\pi(N); \pi$
O_2^+ (or NO)	O_2	O_2 (or OF^+ or NO^-)
$+ \left[:O \equiv O \right]$ e^-	$\longrightarrow$ $:O \equiv O$	$\longrightarrow$ $O \cdot\!\!- O$
sp sp^2	sp sp^2	sp^2 sp^2
$\pi(O); \pi$		$\pi(O)$ $\pi(O)$
OF^+		OF (or F_2^+)
$+ \left[O \cdot\!\!- F \right]$ e^-	$\longrightarrow$ $O \cdot\!\!- F$	$\longrightarrow$ $O \cdot\!\!- F$
sp^2 sp^2	sp^2 sp^2	sp^2 sp^3
		$\pi(O)_{yx}$
F_2^+ (or OF)	F_2	
$+ \left[F \cdot\!\!- F \right]$ e^-	$\longrightarrow$ $F \cdot\!\!- F$ e^-	$\longrightarrow$ F^0 AND F^-
sp^2 sp^3	sp^3 sp^3	Atomic: 3 p orbitals

IV.   PHOTOGRAPHS OF NO AND O₂ MODELS

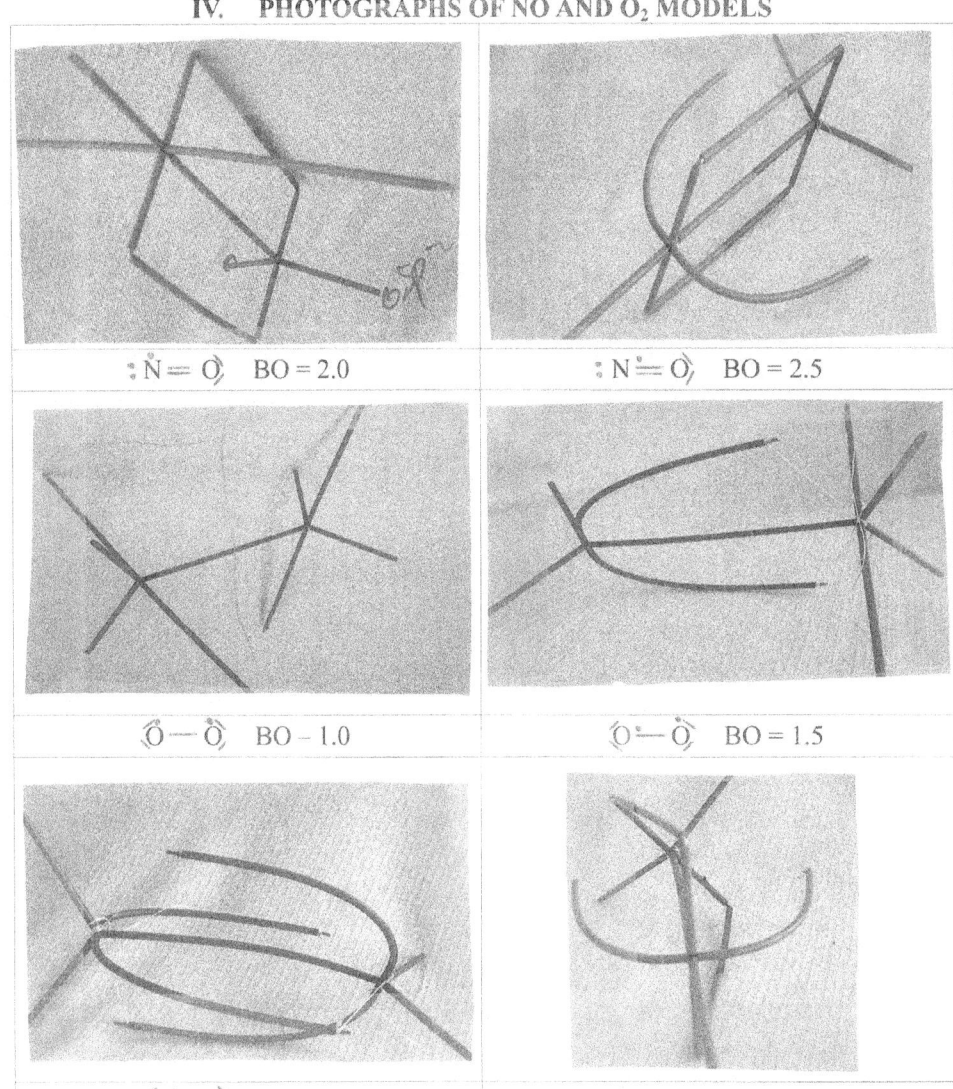

$: \overset{\cdot}{N} = \overset{\cdot}{O}$ BO = 2.0

$: N \overset{\cdot}{=} \overset{\cdot}{O}$ BO = 2.5

$\overset{\cdot}{O} - \overset{\cdot}{O}$ BO – 1.0

$\overset{\cdot}{O} \overset{\cdot}{-} \overset{\cdot}{O}$ BO = 1.5

$\overset{\cdot}{O} - \overset{\cdot}{O}$ BO = 2.0

NO (Skewed)

V. MRAE-MO* ENERGY-LEVEL DIAGRAMS:

LiBe, Be₂, BeB, B₂, BC, C₂, CN, N₂, NO, O₂, OF, AND F₂.
(*Most Recently Added Electron-Molecular Orbital)

One of the great achievements of science is the Aufbau Model for electron filling for atoms. In Section IV is a template for the MRAE-AUFBAU model for periodic elements, which shows not only the Aufbau electronic structures but also exceptions. These exceptions are based *mostly* on what happens to electronic structures when the last, or most recent electron is added to a previous elemental electronic configuration. It appears that a movement of electrons between atomic sub-shells is related to the most favorable energy state for the sub-shell energies, which is based on a sub-shell favoring a half-full or full complement of electrons.

The MRAE-MO diagrams that follow is a similarly based system. For example, on addition of an electron (MRAE) to an orbital labeled pi* (anti-bonding), a shift of an electron from a pi (π) electron pair into this pi* orbital occurs.

LiBe; Li:: $2s^1$; Be: $2s^2$	Be₂.	BeB; B: $2s^2 2p^1$	B₂
------- -------π_b	------- -------	------- ↓	↓ ↓
------- ------σ_{nb} ↓	------- ↑↓	↑↓ ↑↓	↑↓ ↑↓
⇗ ⇙↗ ↑↓ ------σ_b	⇗ ⇙↗ ↑↓	⇗ ⇙↗ ------	---------
BO = 0.75[1] P = 1	BO = 0.5[2] P = 0	BO = 0.75[3] P = 1	BO = 1.0[3] P = 2
CLi — Be:	CBe — Be:	:Be ∴ B:	⦂B ⦂ B⦂

[1] The BO for LiBe is estimated as 0.75 as the MRAE goes into one of the equal-energy σ_{nb} regions, which can pair with one of the σ bonding electrons or apply pressure on the bonding pair to move into the second (vacant) σ non-bonding region. Also, its BO is between Li₂ and Be₂.

[2] Be is a metal, and Be₂ is not similar to 'He₂'. Consequently, a bonding pair is estimated to spend one-half time in the bonding region, explaining the estimated BO of 0.5.

[3] It appears, based on known BO values, that pi orbital electrons can 'perturb' into the sigma bonding region. This tends to keep, totally or partially, the sigma electron(s) in the non-bonding sigma regions.

BC; C: $2s^2 2p^1 2p^1$	C_2 (GS)[1]	CN; N: $2s^2 2p^1 2p^1 2p^1$	N_2
―― ―π* ―― ――	―― ―― ―― ――	―― ―― ―― ――	―― ―― ―― ――
↓ ↑↓―π	↑↓ ↑↓	↑↓ ↑↓	↑↓ ↑↓
↑↓ ↑↓―σ_{nb}	↑↓ ↑↓	↑↓ ↑↓	↑↓ ↑↓
――σ_b	――	↑↓	↑↓
BO = 1.5 P = 1	BO = 2.0 P = 0 [C_2 (ES) P = 2]	BO = 2.5 P = 1	BO = 3.0 P = 0

⦿B ― C⦿	⦿C = C⦿	⦿C ⩵ N⦿	⦿N ☰ N⦿
			HC☰CH sp sp

[1]This is ground state (25^0C) C_2. An excited state form at 7.3 KJ higher energy has been characterized: BO = 2.0, P = 2.

⦿ C ⩵ C⦿

[NO]: O: $2s^2 2p^2 2p^1 2p^1$	NO^2	$[O_2]$	$O_2\,^3$

(molecular orbital diagrams with electron arrows; labels: π^*, π, σ_{nb}, σ for column 1; (N), sp^2 for column 2; (O) for column 3; (O), (O), sp^2 for column 4)

BO = 2.0 or 2.5 $P = 1$	BO = 2.5 $P = 1$	BO 1.0 or 2.0 $P = 2$	BO = 2.0 $P = 2$
$:\!\ddot{N} = O\!:$ or $\cdot\ddot{N} = \ddot{O}\!:$	$:N = O$ or $:N \doteq O$	$[\;:\!\ddot{O} - \ddot{O}\!:\;$ or $:\!\ddot{O} \doteq \ddot{O}\!:\;]$	$O = O$ or $\langle O \cdots O \rangle$

BO=2.0; P=0
$H_2C = CH_2$
$O = O$ (ES)
$HN = NH$

[2] σ_{nb} and p^* of oxygen hybridize to sp^2. *Lone p orbital pi bond* forms.

[3] σ_{nb} and p^* second oxygen hybridize to sp^2. *Lone p orbital pi bond* forms. (ES O_2 does exist and some plants synthesize it first as it does have a function.)

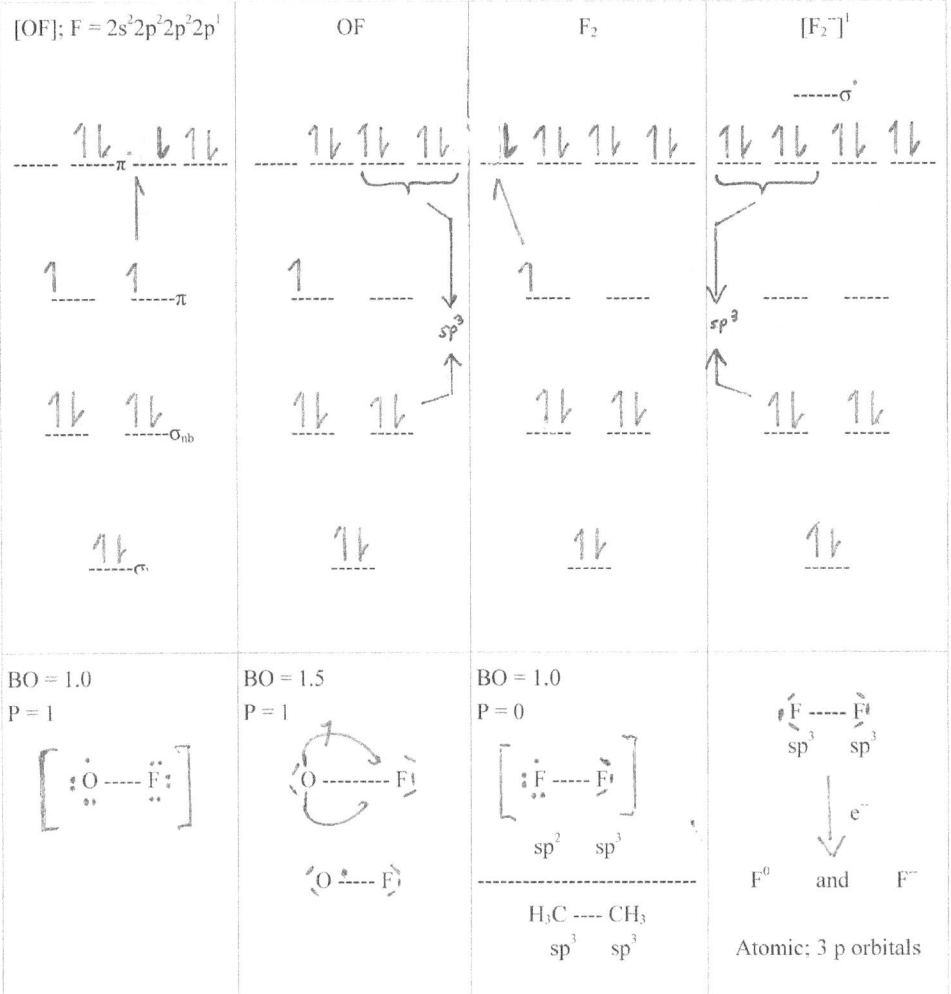

[OF]; F = $2s^2 2p^2 2p^2 2p^1$	OF	F_2	$[F_2^-]^1$
BO = 1.0 P = 1	BO = 1.5 P = 1	BO = 1.0 P = 0	

[1]This species fragments into a fluorine atom and a fluoride ion.

VI. MRAE-MO* ENERGY DIAGRAMS TEMPLATE

*Most Recently Added Electron-Molecular Orbital

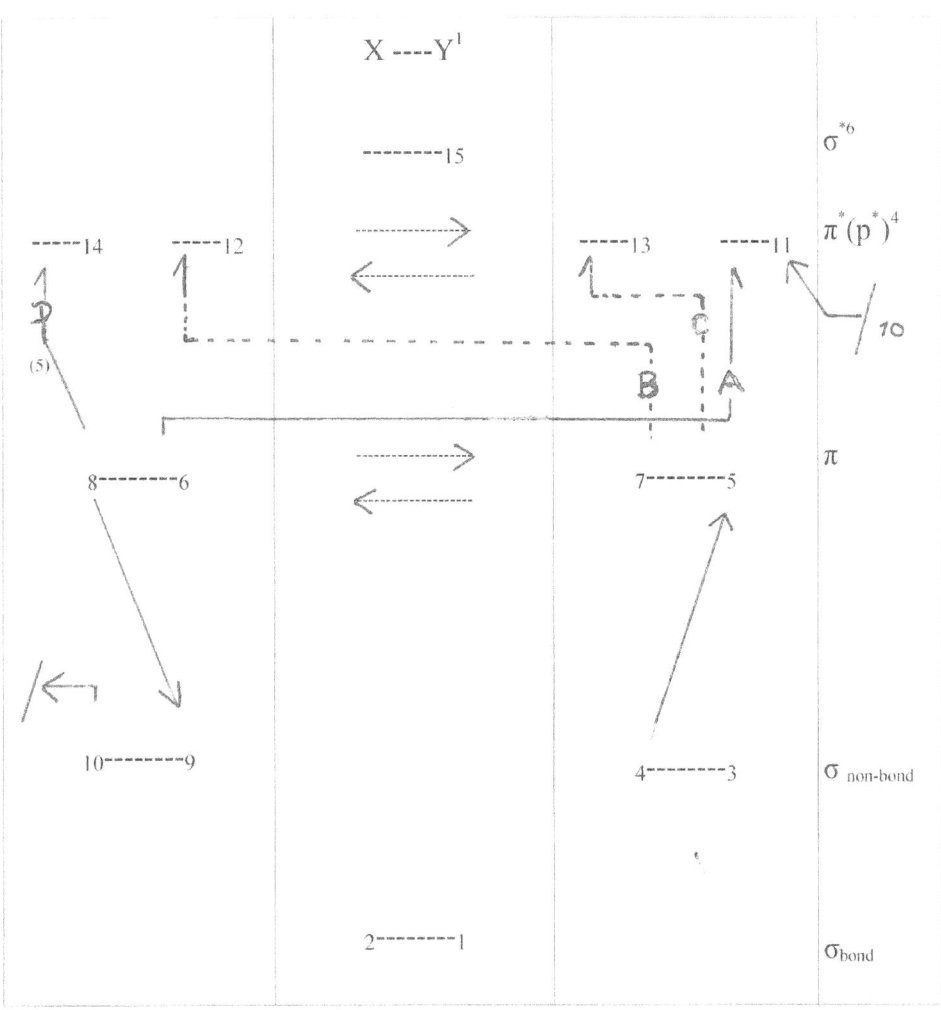

$X ---- Y^1$

--------15

σ^{*6}

------14   ------12   ------13   ------11 $\pi^*(p^*)^4$

D

(5)

A C B

/ 10

π

8--------6   7--------5

10--------9   4--------3 $\sigma_{non-bond}$

2--------1 σ_{bond}

[1] Left to right, row two Periodic Table.

[2] → Electron filling order.

[3] ↝ Movement of a single electron between orbitals.

[4] MRAE (electron) is anti bonding as to π orbital electron pair, and electron remaining is a *Lone p Orbital pi* (LPOP) electron.

[5] The π orbitals are now void of electron density; p (p*) orbitals and/or hybrid orbitals remain.

[6] An electron entering this orbital causes fragmentation of the σ bond. F_2 would fragment into F^0 and F^-.

CHAPTER TWO

PART I.   POLYATOMIC STRUCTURES AND SEQUENCES
PART II. MECHANISTIC HYBRIDIZATION ARTICLE

PART I

I.   Hybridization And VSEPR Revisited. In the 1987 version of AMEC the main focus related to arriving at molecular structures through the method of Linear Combination of Atomic Orbitals (LCAO). The emphasis was that by recognizing the character of the di-lobed p orbital and applying the principles of hybridization, which incorporates VSEPR (Valence Shell Electron Pair Repulsion) theory, that most, if not all, structures involving row two elements are explained. The hybridization guiding principles at that time, and more so today, can be defined as follows:

a. Bonding: If a second atom (H, etc.) bonds to an atom, usually at one of the three equal energy p orbitals, then the *original* atomic structure no longer exists, at least not in its original form.

When the added atom bonds at a specific p orbital lobe electron density becomes concentrated in the newly created *sigma (σ)* **bonding region (symmetric about the x axis passing through the two bonding atomic cores).** Consequently, the opposite side p lobe becomes largely devoid of electron density, producing a *small back lobe*.

b. **Hybridization.** For Row Two elements, an added atom can produce three levels of hybridization depending upon the hybridization level of the original species. The sequence, as seen in Chapter 1, is: atomic > sp > sp^2 > sp^3. Basic shapes as related to hybridization are as follows: sp, linear; sp^2, trigonal; sp^3, tetrahedral. However, only the *saturated* sp^3 hybridized forms are rigid in structure, examples being: BH_4^-, CH_4, NH_4^+, OR_4^{+2} (and possibly, FR_4^{+3}).   For the hybridization sequences that follow carbon and nitrogen sequences will be demonstrated.

1. **sp Hybridization (linear distribution, two p orbitals remain).** When an atom bonds to an atomic state Row Two (Periodic Table) element, at a p lobe, two things occur. First, the bonding electron pair must be concentrated between these two atoms, forming a 'covalent' sigma bond. Second, the spherical 2s electron density becomes distorted as it is 'shielded' to a degree from the region that now comprises the sigma covalent bond region.

The result being that neither the atomic 2s orbital (spherical) nor the atomic 2p orbital (two equal lobes) now exists. The p lobe opposite the bonding region becomes a small back lobe since the covalent bonding electron pair forms a sigma (σ) bond region. The p orbital 'small back' lobe region allows the 2s electron density to occupy more of this region.

At this point, the electron distribution of this 'sp' mixing or sp hybridized structure is not greatly different from the electronic distribution of the original atomic structure. The 2s spherical orbital is only marginally modified and the two sp hybrid (σ) bonding orbitals are oriented along the p orbital axis. In addition, two of the original three p orbitals remain largely, if not totally, unaltered.

2. sp² Hybridization (trigonal distribution, one p orbital remains). sp² and sp³ hybridization are very different from sp hybridization as there is no 2s (spherical) type electron density to occupy a small back lobe region. Consequently, when an atom bonds at one p lobe creating a small back lobe region, an unstable arrangement of electron pair repulsion occurs. (Alternatively, a p electron pair can shift into one lobe region creating a small back lobe.) The result is that realignment of electron pair densities occurs: the sp linear arrangement becomes an sp² trigonal arrangement invoking the VSEPR rule. Three *equal* sp² hybrid regions orient at an angle of 120⁰.

3. sp³ Hybridization (tetrahedral distribution, zero p orbitals remain). The same sequence as above occurs when an atom bonds to the last p orbital of an sp^2 hybrid form, or an electron pair concentrates in one lobe (of a p orbital). If the four sp^3 regions are identical, the angle between the core atom and two attached groups is 109.5°. However, if one region is occupied by and electron pair, which is more spherical ('fatter') than a bonded pair, the other groups are compressed, the degree being related to the relative sizes of the groups.

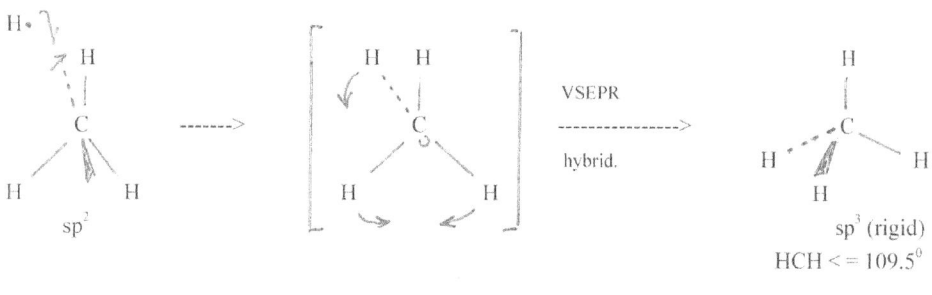

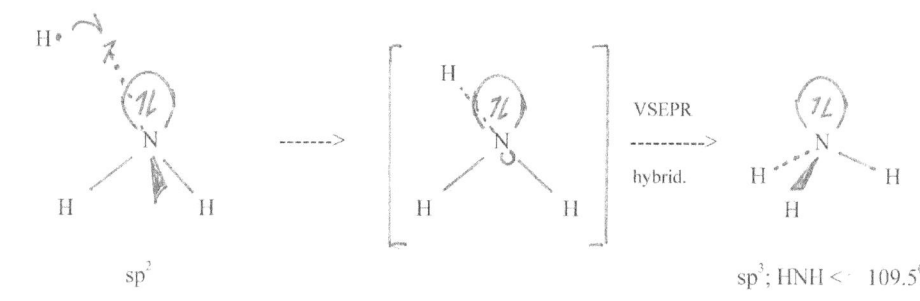

NH₃ **Inversion of Configuration and Rigid NH₄⁺.**

PART II

**EDUCATION IN SIBERIA
INTERNATIONAL EXPERIENCE AND
COOPERATION IN EDUCATION,**
No. 13, 2005. pp 97-100

MECHANISTI HYBIDIZATION

Harold J. Teague

University of North Carolina at Pembroke, USA

The concept of mechanistic hybridization was presented at the 10th Biennial Conference on Chemical Education, Purdue University, 1988: *A Construction Method for Polyatomic Structures* and in a self-published work, 1987, *Atomic and Molecular Electronic Configuration.*[1]

Quantum mechanics shows that the carbon atom (and also other row two elements, particularly nitrogen and oxygen) in organic molecules exists largely in an sp[3]

---

[1] The concept was presented at numerous local and regional workshops and conferences as well.

hybridized state, an sp^2 hybridized state or in an sp hybridized state. Also, due to resonance and other factors immediate hybridization states can, and do, exist.

What is hybridization? A stand-alone carbon atom, C is known to possess one valence shell spherical orbital, labeled 2s and three 2p orbitals. A p orbital is non-bonded and di-lobed with the identical lobes oriented $180°$ to each other and symmetric to the central atom core. The three p orbitals of a carbon atom are at right angles to each other and are often presented as oriented along x, y and z axes. In the ground state carbon atom two paired electrons are presumed to be in the spherical 2s orbital, one electron each in two of the 2p orbitals and one p orbital void of electron density. The electron configuration can be represented as: C: $[k]2s^2 2p_x^1 2p_y^1 2p_z^0$. In an excited carbon atom one 2s electron has moved into the vacant p orbital: C^*: $[k]2s^1 2p_x^1 2p_y^1 2p_z^1$.

In sp^3 hybridization there are four hybrid (sp^3) orbitals arising from hybridization (mixing) of the 2s orbital and the three 2p orbitals. The four hybrid sp^3 orbitals that replace the one 2s and three p orbitals form a tetrahedral arrangement, i.e. each hybrid orbital is directed towards the corner of a tetrahedron. In sp^2 hybridization the 2s orbital and two of the 2p orbitals hybridize (mix) to produce three sp^2 hybrid orbitals that form a trigonal planar arrangement. One of the p obitals remains essentially unaltered. In sp hybridization the 2s orbital and one p orbital hybridize (mix) to produce two sp hybrid orbitals that form a linear arrangement. Two p orbitals remain essentially unaltered (1,2).

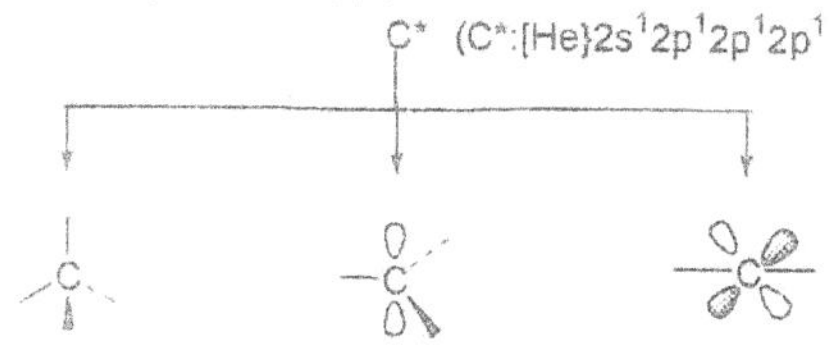

$$C^* \quad (C^*:[He]2s^1 2p^1 2p^1 2p^1)$$

Scheme 1

This article focuses on how hybridization relates to structure as one species undergoes transformation into a second species. The process is pathway, or mechanism, based. Mechanistic hybridization involves two basic steps.

A. A species, usually a H atom, is proposed as reacting with the starting species through a specific orbital, which occurs usually at one lobe of a p orbital. This results in a small back lobe where the original non-bonding larger lobe existed.[1]

B. Hybridization, i.e. mixing of orbitals, follows. The overall result is a re-distribution of electron regions and atoms in accordance with VSEPR rules to form a

---

[1] Not only can the small back lobe concept be visualized qualitatively as well as inferred from the facts of organic chemistry but has also been calculated by quantum mechanical methods (2).

new hybrid structure, i.e. a structure possessing maximum bond energy.

As an example suppose the H --- X --- H species is linear with one electron in a p orbital (perpendicular to bond axis) and that a H atom initiates bonding at one lobe of the p orbital. As the bond further forms the one electron in the p orbital and the electron of the H concentrate in the bonding area, i.e. form a covalent bond region between the X and the H. This produces a small back lobe where the larger p lobe was before bonding. At this point it is convenient to picture that the three H's form an unstable "T" distribution around the central X atom, i.e. non VSEPR arrangement.

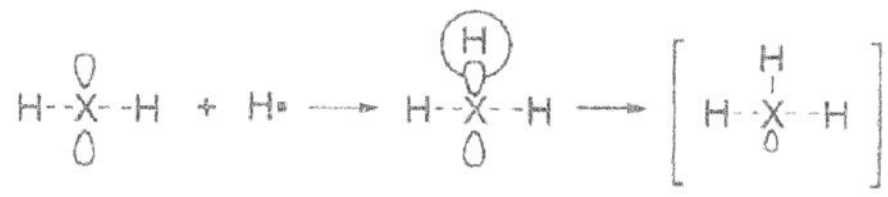

Scheme 2

The "T" shaped intermediate species is of high energy (unstable) because of the unequal distribution of the three valence shell electron pair regions, non VSEPR, which include the H's around the central X atom. A re-distribution occurs in accordance with VSEPR rules to the more stable trigonal structure, yielding maximum bond energies. Hybridization to a higher state, e.g. sp -> sp^2, etc. has also occurred as a p orbital has been merged into the overall system. (Note. Any non-bonded p orbitals remain largely unaltered.)

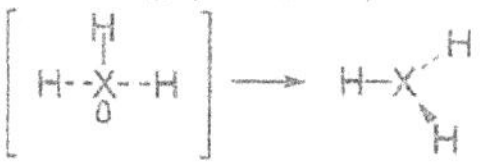

Scheme 3

From the above example it follows that a key component of the mechanistic hybridization process is the effect produced in a p orbital when an atom bonds to it.

The basis for study of mechanistic hybridization will be two interrelated sequential systems: $C + 2H_2 \rightarrow CH_4$ and $2C + 3H_2 \rightarrow CH_3CH_3$. In the mechanistic hybridization scheme below a hydrogen atom is shown as the bonding species.

I.   sp Hybridization.   kcal/mol

a. $C \rightarrow C^*$; $C^* + H \rightarrow CH$ $\Delta H = +90$ (calc.[2])

b. $CH + H \rightarrow HCH$ (linear) $\Delta H = -102$ (3)

c. $2CH \rightarrow HCCH$ $\Delta H = -230$ (3)

II. sp^2 Hybridization.

d. $CH_2 + H \rightarrow CH_3$ $\Delta H = -110$ (3)

e. $HCCH + H \rightarrow H_2CCH + H \rightarrow H_2CCH_2$ $\Delta H = -42$ (4)

---

[2] Calculated from bond energy data and heat of formation data (3) by summation of the following:

CH_4	$\rightarrow CH_3 + H$	+104 kcal/mol
CH_3	$\rightarrow CH_2 + H$	+110
CH_2	$\rightarrow CH + H$	+102
$C + 2H_2$	$\rightarrow CH_4$	-18
$4H$	$\rightarrow 2H_2$	-208 (Note. The

calculation uses ground state carbon.)

III.   sp³ Hybridization.
    f. $CH_3 + H \rightarrow CH_4$ $\Delta H = -104$ (3)

    g. $H_2CCH_2 + H \rightarrow H_3CCH_2 + H \rightarrow H_3CCH_3$ $\Delta H = -33$ (4)

    sp Hybridization.

I.a. $C^* + H \rightarrow CH$. Although the CH species, neither formation nor degradation, has apparently been studied, its formation can be pictured as follows: An atom (H) associates with the bonding axis p orbital of an excited carbon atom, C^*. Electron density of both the p orbital and the H shift into the bonding region, which results in a temporary decrease of electron density in the back lobe area. The spherical electron density of the 2s orbital becomes less spherical (perturbed) by occupying more of the back lobe (non-bonding) region along the bonding axis. The result can be described as equating to sp hybridization, i.e. two sp hybrid regions exist along bonding axis and two p orbitals remain largely unaltered.

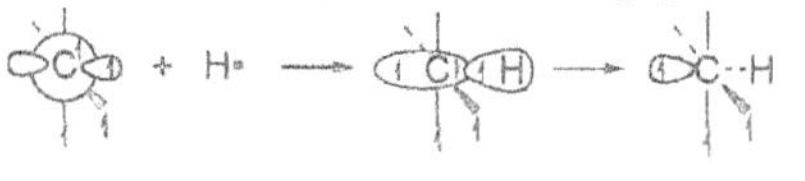

Scheme 4

I.b. $CH + H \rightarrow HCH$ (linear). An H atom associates with, and then bonds, to the non-bonded sp orbital, yielding two identical CH bonds along with the largely unaltered two p orbitals.

Scheme 5

I.c. $2 CH \rightarrow HCCH$. Two CH species are pictured as bonding together through one bond (termed sigma) originating form two sp orbitals and two bonds (termed pi) from the two p orbitals of each CH species. The result is a triple bond between the two C's.

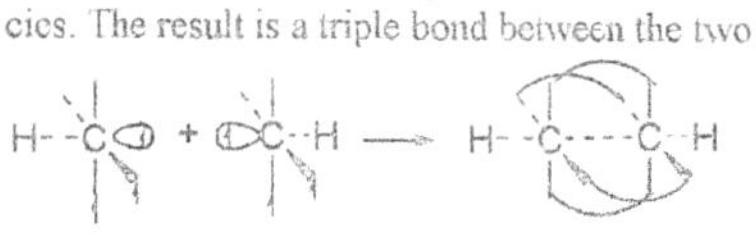

Scheme 6

sp² Hybridization.

II.d. $CH_2 + H \rightarrow CH_3$. An atom (H) is proposed as associating and bonding to one of p orbitals of CH_2. The p orbital and H electron density concentrate in the bonding region, leaving a small back lobe. An unstable "T" shaped CH_3 species is the result. Unlike in the C^* atom there is no 2s (spherical) electron density available to move into the back lobe region. The H atoms shift from a "T" distribution to a favored trigonal arrangement, in accordance with VSEPR rules. One p orbital remains as the system is now in an sp² hybridized state.

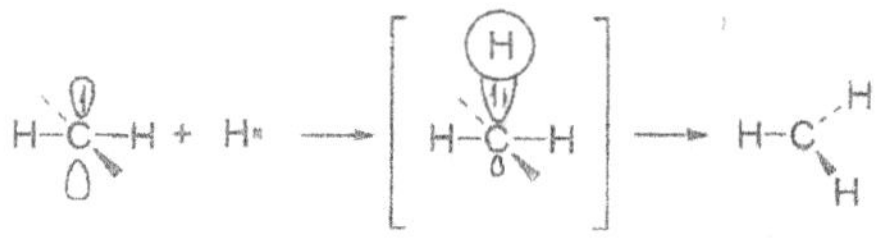

Scheme 7

II.e. $HCCH + H \rightarrow H_2CCH + H \rightarrow H_2CCH_2$. In the presented scheme one pi bond of HCCH is shown as being broken, i.e. p orbitals re-form. An atom (H) associates and bonds to the p orbital of one of the carbons, leaving a small back lobe and a "T" shape as seen previously. Rearrangement to a trigonal distribution (sp² hybridization) occurs. The same sequence is envisioned at the second CH to yield planar, trigonal $H_2C = CH_2$.

It should be noted that although a step process, for illustration, is presented here, a concerted (same side) catalyzed addition of the two H atoms is the more practical pathway.

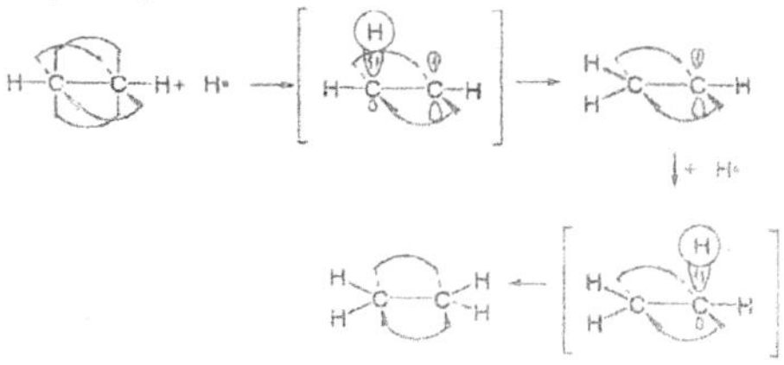

Scheme 8

sp³ Hybridization.

III. f. $CH_3 + H \rightarrow CH_4$. An atom (H) is shown as bonding to the 'last' p orbital of trigonal CH_3. As previously, the H bonds and a small back lobe forms to forme an unstable trigonal pyramidal shape. A re-distribution of the H atoms occurs to form the favored, tetrahedral CH_4 that is sp³ hybridized. The spherical 2s orbital and the three 2p orbitals have all 'merged' to form four sp³ hybrid orbitals and yield the perfectly symmetrical CH_4 molecule.

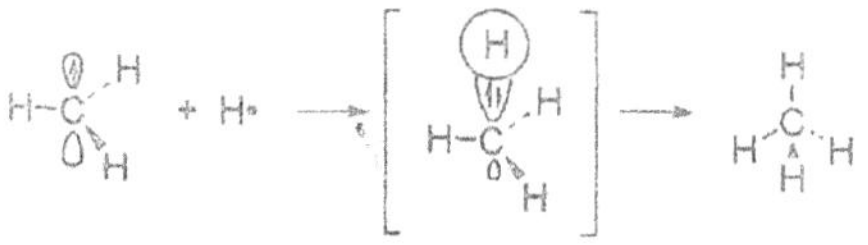

Scheme 9

III. g. $H_2CCH_2 + H \rightarrow H_3CCH_2 + H \rightarrow H_3CCH_3$. As in III. F. an atom (H) bonds to the p orbital of one of the carbons producing a small back lobe and an unstable trigonal pyramidal shpape. A re-distribution occurs to form a tetrahedral structure at that carbon. The same process is repeated at the second carbon. The result is tetrahedral H_3CCH_3.

99

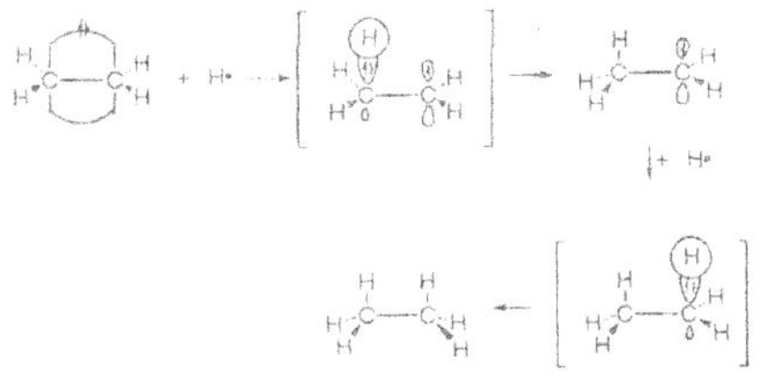

[Scheme 10 diagram]

Scheme 10

In the above examples an atom bonding to a p orbital initiates hybridization, i.e. the p orbital becomes a bonded hybrid orbital. It should be noted that an atom does not need to bond to a p orbital for hybridization to occur, only a shift in electron density from a di-lobed p orbital into a hybrid orbital. One example is the conversion of CH_2 (linear) into CH_2 (trigonal). One pathway can be presented as follows. An electron in one p orbital moves into the second p orbital giving that p orbital an electron pair. The electron pair becomes largely localized in one p lobe region while the second lobe region becomes a small back lobe. This process produces an unstable "T" shape distribution of the two H atoms and the non-bonded electron pair. Re-distribution to a stable trigonal arrangement (sp^2 hybridization) follows.

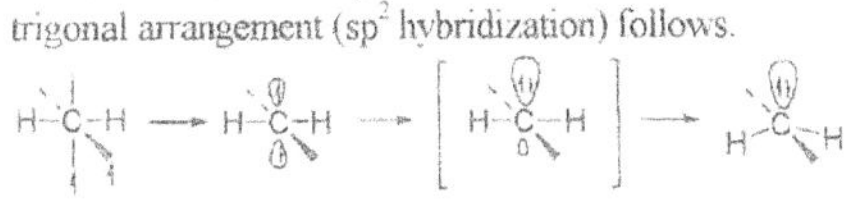

[Scheme 11 diagram]

Scheme 11

Another example is the non-rigid NH_3 molecule (also CH_3^- and H_3O^+). The lowest energy form of NH_3 is sp^3 hybridized; however, there are two equivalent sp^3 forms, each form possessing a non-bonded electron pair in an sp^3 hybrid orbital with a small back lobe. One sp^3 form undergoes 'inversion' into the second, equivalent form. In the process of doing so the NH_3 molecule passes through a high energy intermediate sp^2 hybridized state where the non-bonded pair is in a p orbital and the N and three H atoms are in a planar trigonal arrangement.

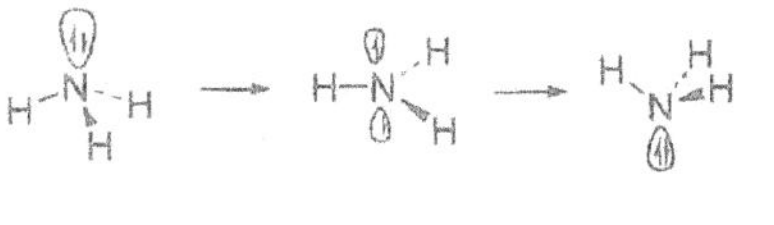

[Scheme 12 diagram]

Scheme 12

If a molecule containing a nitrogen (A) has rigid geometry because the groups attached to the N atom are locked in position due to bridging, or other reason, then inversion cannot take place, nor can a different level of hybridization occur. There exists a firm correlation between hybridization and the geometry of the atoms around a central atom, not the number of atoms. For instance, the NH_3 molecule is sp^3 hybridized and has H-N-H angles of about $107°$; not the exact angles ($109°28'$) of a perfect tetrahedral structure such as CH_4 but close. Another example is cyclopropane (B). Although each carbon has four atoms bonded to it, the molecule possesses much ring strain, i.e. the C-C bonds are not at maximum energy, as the C-C-C angles are $60°$ (and the H-C-H angles are greater than the normal sp^3 distribution). Consequently, the hybridization of the carbon atoms is somewhere between sp^3 and sp^2.

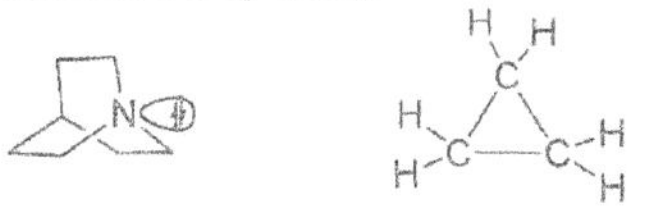

[Structures A and B diagram]

A   B

Scheme 13

Summary. Mechanistic hybridization is a system that this author has found over twenty years to be a useful tool in teaching how structure and hybridization is related. It is intended to supplement the traditional approach to hybridization not replace it. Students grasp the relationship between hybridization and the number of p orbitals and are able to appreciate how atom carbon can be the central atom of a sp^3 (tetrahedral) hybridized system.

REFERENCE

1. Pauling, L. *The Nature of the Chemical Bond*, 3rd ed.; Cornell University Press: Ithaca, NY, 1960; Chapter 4. *The Nature of the Chemical Bond: Application of Results Obtained from the Quantum mechanics and from a Theory of Paramagnetic Susceptibility to the Structure of Molecules*, J.A.C.S. 53, 1367, 1931.
2. McMurry, J. *Organic Chemistry*, 3rd ed.; Brooks/Cole: Pacific Grove, CA, 1992; Chapter 1.
3. *Handbook of Chemistry and Physics*, CRC Press, 1975; D-82 and F-224.
4. Fessenden, R.; Fessenden, J.; Logue, M. *Organic Chemistry*, 6th ed.; Brooks/Cole: Pacific Grove, CA, 1998; Chapter 2.

CHAPTER THREE

SELECTIONS FROM

ATOMIC AND MOLECULAR
ELECTRONIC CONFIGURATION
1987©

MRAE*-AUFBAU TEMPLATE

*MOST RECENTLY ADDED ELECTRON.

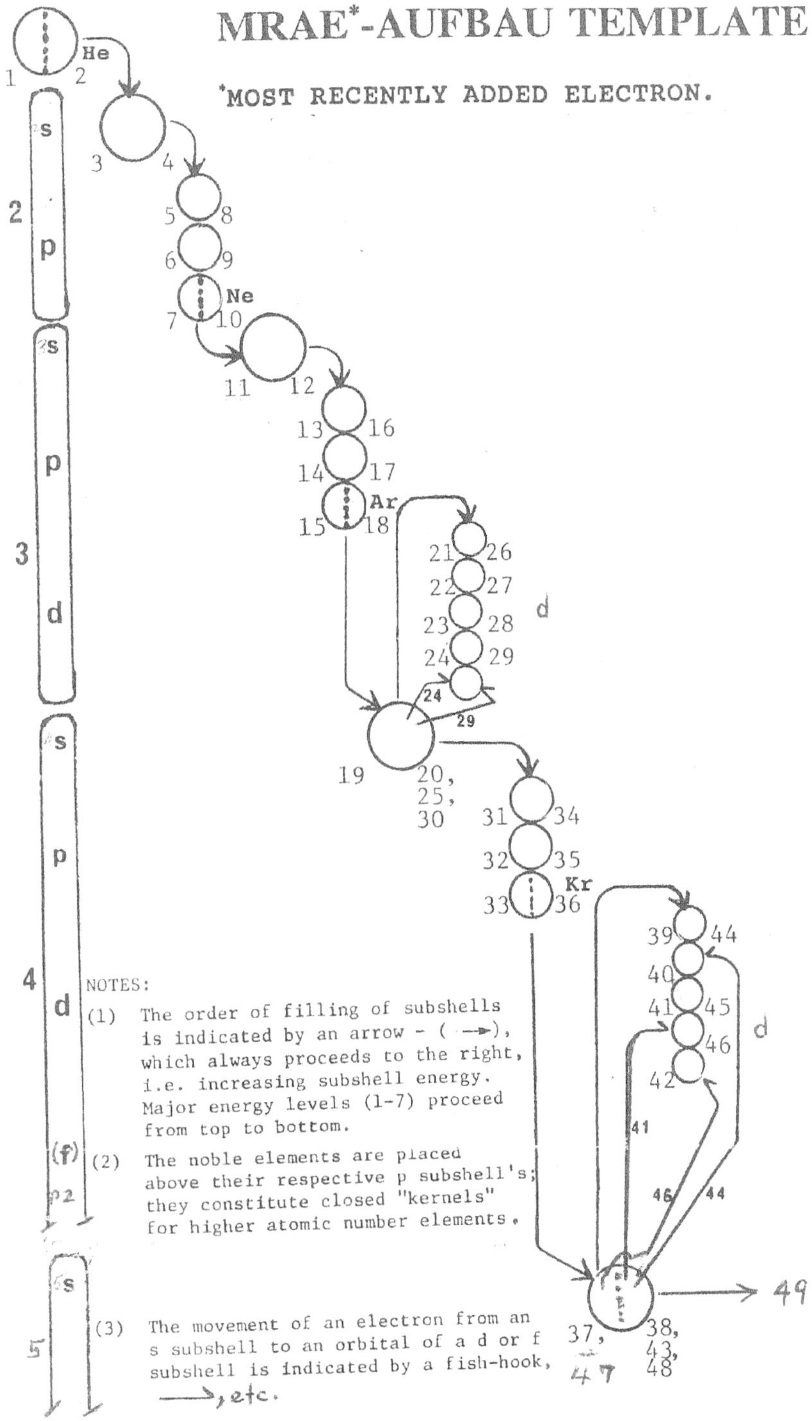

He

Ne

Ar

d

Kr

d

NOTES:

(1) The order of filling of subshells is indicated by an arrow - (→), which always proceeds to the right, i.e. increasing subshell energy. Major energy levels (1-7) proceed from top to bottom.

(2) The noble elements are placed above their respective p subshell's; they constitute closed "kernels" for higher atomic number elements.

(3) The movement of an electron from an s subshell to an orbital of a d or f subshell is indicated by a fish-hook, →, etc.

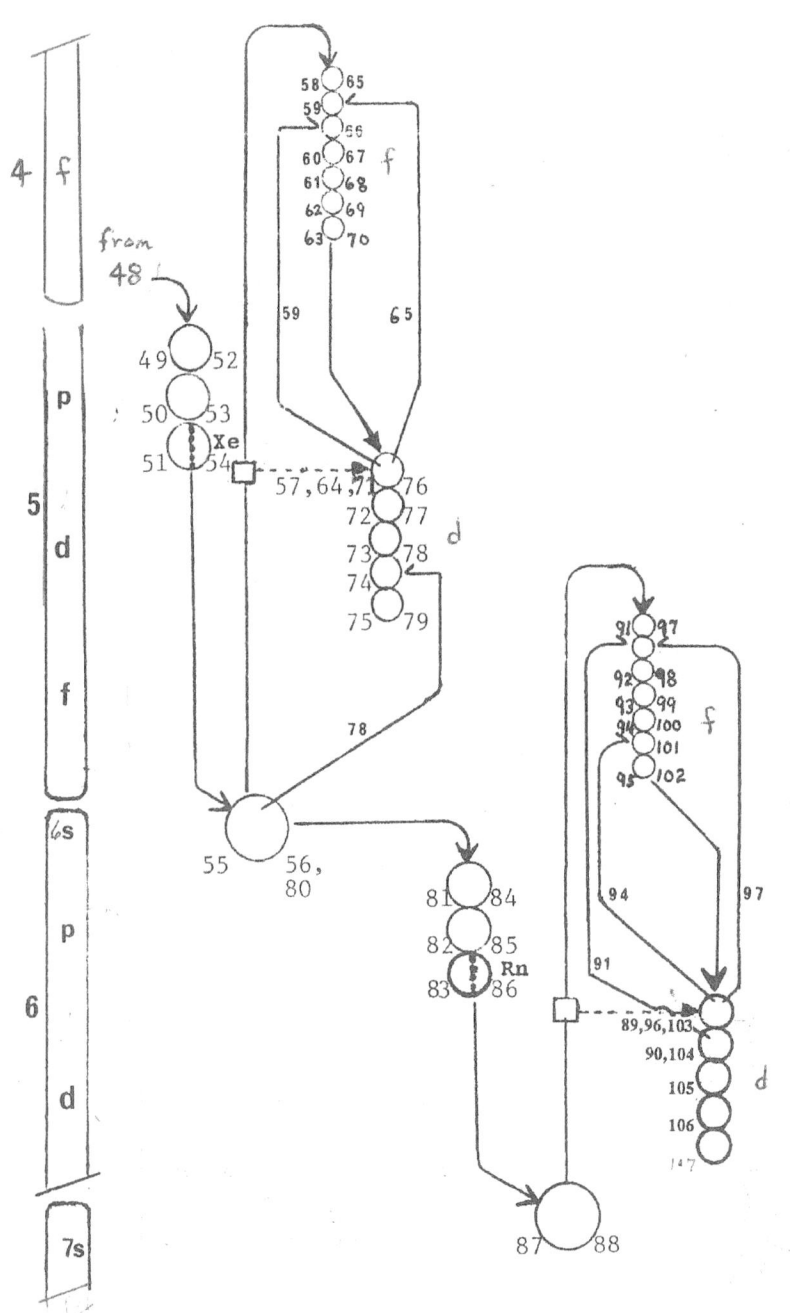

NUCLEUS-ELECTRON INTERACTION

Assume an isolated electron is moving in a "straight" line
at a constant velocity and that it comes under the influence
of a positively charged body (proton), Figure 1.4. What would
be the result? The electron moving in a wave pattern (spiraling
either right or left-handed) would be pulled toward the nucleus.
As the electron moves toward the nucleus it is accelerated
(and has a corresponding change in wavelength); however, instead
of being pulled into the nucleus the momentum of the electron
causes it to "sling" around the nucleus, i.e. into an orbital
pattern, Figure 1.4.

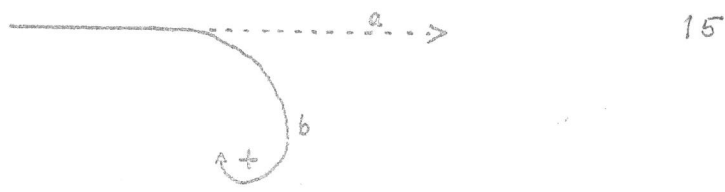

Fig. 1.4. The electron, moving along path a, comes under the
influence of the nucleus, and is accelerated towards the
nucleus (+). Path b shows resultant, i.e. movement of electron
as it moves around nucleus.

The electron should carve out a single
circuit similar to that depicted in Figure 1.5,
although the exact path can never be known
(Uncertainty Principle). The electron in region
P will have low velocity and high potential
energy (in relation to the nucleus (+)). In

Fig. 1.5

region A the electron will be accelerating, to some fraction of
the speed of light, whereas in region D the electron will be
de-accelerating.

Why would one propose an elliptical orbit? The mass of the proton is 1837 times that of the electron, and both are appreciably, and oppositely, charged. When both factors, relative masses and charges, are considered an elliptical path is reasonably expected. An appropriate analogy is the elliptical path of Halley's Comet.

Does the electron stay in the exact same pattern around the nucleus? If the energy of the electron (and atom) stayed exactly constant, then it is theoretically possible that a single path (i.e. orbit) would result. However, the electron can absorb and emit energy (change direction, velocity, wavelength, etc.) from its interaction with its environment (low energy electro-magnetic radiation, etc.), and the path of the electron is altered, if only slightly, with each passage around the nucleus! (These changes are slight perturbations, as opposed to photon excitations, and the electron stays in the same energy level. Photon excitation can cause promotion of an electron to a higher energy level orbital, as discussed below.)

The result of these altered paths, over time, with each passage around the nucleus leads to a model of hydrogen as depicted in Figure 1.6.

The main features of the model are as follows:

1) A sphere would be carved out over time by movement of an electron around a nucleus.

Fig 1.6

2) Total volume, excluding nuclear area, would contain electron density, i.e. all space in atomic volume referred to as orbital) would contain the electron at

one time or another.

3) Some regions of atomic volume would have greater
probability of containing electrons at any given time
than other regions.

Analogy. The appropriate analogy for the ground state
hydrogen atom is the path carved out by Halley's Comet over an
extremely large number of passes around the sun. If Halley's
Comet could be plotted over a million/billion years the result
would be similar to the electron density image of a hydrogen atom
generated in a second, or less.

To visualize the electron density probability plot of a
hydrogen atom think of Halley's Comet hypothetically plotted
over a million year period. We know that: 1) one cycle of
Halley's Comet requires about 83 years, 2) that it spends
very little time in the vicinity of the sun due to its great
velocity as it approaches the sun, and 3) that it is moving
relatively slow at its extremities, which correspond to vast
volumes of space. Now, suppose you have a space ship and a
single day during this one million year period to return to the
solar system to attempt to locate Halley's Comet! Where would
you look? Before you decide, work out a probability plot of
where Halley's Comet is likely to be at any one time. Does your
plot look anything like Figure 1.7? In Figure 1.8 the shaded
volume corresponds to the volume in which the comet is most
likely to be found! (Slight pertubations alter the path of the
comet on each circuit; therefore, a sphere should be carved
out over millions of years.)

II. MOLECULAR ELECTRONIC SYSTEMS INVOLVING p ATOMIC ORBITALS

Elements five through ten (B,C,N,O,F AND Ne) of the second row have electrons in p atomic orbitals. The bonding situation for these elements is different from previous systems since p orbital overlap is also involved. Molecular species involving these elements will be considered in two groups. The first group (A) contains systems that have from six to ten total valence electrons; the ten electron system corresponds to N_2, the most stable two-nuclear molecule. The second group (B) consists of molecular species that have greater than ten valence electrons.

A. Dinuclear Systems With Ten or Fewer Valence Shell Electrons

4. B_2 (Diatomic Boron)

Boron has three valence electrons, two in the 2s orbital and a third in one of the p orbitals. Two other p orbital areas exist that are vacant of electron density. Two boron atoms can interact with each other along the x-axis as illustrated below. In this representation the x-axis p orbital is vacant, or non-existant, while an electron is in a y-axis p orbital of one atom and a z-axis p orbital of the other. The p orbitals overlap

side-to-side to form two degenerate pi(π) bonding molecular
orbitals, each possessing a single electron (Hund's Rule is obeyed).

Two pair of electrons are in the sigma (σ) orbitals. The
sigma orbitals can be viewed as a bonding/anti-bonding combination,
or as equivalent non-bonding components. Regardless of how these
sigma electron pairs are viewed they effectively cancel each other
and are not seen as effecting bond strength (see helium system).

Electron density is "resonating" in and out of the sigma
bonding area. Although this σ electron density does not contribute
appreciably, if at all, to bonding it does keep p orbital electrons
from going into this area. Consequently, bonding is a result
of π orbital density, not sigma bonding density!

The electronic configuration system for B_2 is constructed as
follows (remember only one component of each π orbital is shown
and the z-plane pi component is shown "skewed").

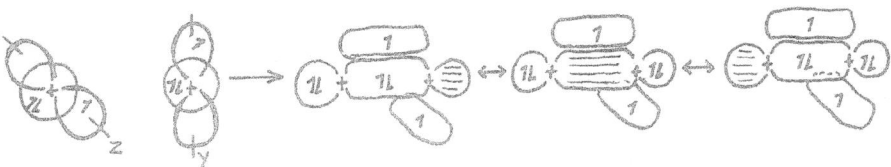

The B_2 molecule is known to have a bond order of one and to
have two unpaired electrons. This electronic configuration is in
agreement with the experimental data for B_2; it is favored because
minimal alteration in atomic configuration is required (for
example, no change in the p orbitals is necessary except for
side-to-side overlap).

Electron-dot Structure. $:B-B:$ or $:B \overset{|}{\cdot} B:$

5. C$_2$ (Diatomic Carbon)

The same reasoning that was used for B$_2$ can be extended to C$_2$.
Two molecular forms of C$_2$ are known to exist, and electronic
configurations for these C$_2$ forms are constructed below. A
carbon atom has four valence electrons: an electron pair is in
the 2s orbital and unpaired electrons are located in two of the
three p orbitals (Hund's Rule).

a) C$_2$, Ground State. The electronic configuration for the
preferred form of C$_2$ is constructed in the same manner as was B$_2$.
The 2s orbitals interact along the x-axis to form σ orbitals and
the y-axis and z-axis p orbitals overlap side-to-side to form
degenerate π orbitals. This configuration is constructed as
follows:

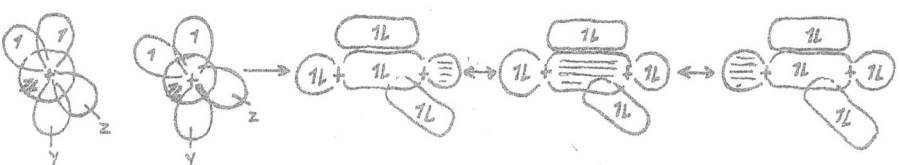

As with previous systems the sigma electrons are "non-bonding";
therefore, C$_2$ with this configuration has a bond order of two
and is diamagnetic; it is consistent with experimental data for
ground state C$_2$.

$$:C = C:$$

b) C$_2$, Excited State. A second molecular form of C$_2$ is known
to exist that is 7.3 KJ per mole higher in energy than the previous
(ground-state) form. It is paramagnetic (has unpaired electrons)

and involves sp hybridization[a] of at least one atom.

   sp Hybridization. The sp type occurs when there exists an
x-axis p orbital and a second nucleus interacts with it along
the x-axis. The second nucleus causes a distortion in the p
orbital lobes (one gets bigger while the other almost disappears)
and the 2s orbital is "pushed" to the backside. This induced distortion
causes a merging and redistribution of electron density in these
two orbitals, i.e. hybridization occurs. The result is the formation
of two identical sp "hybrid" molecular orbitals, occupying
original s and p_x orbital areas. Formation of sp hybrid orbitals can
be depicted as follows:

2s, px orbitals   Interaction with   sp hybridization
                           second nucleus

   The electronic configuration for the high energy C_2 molecule is
as follows:

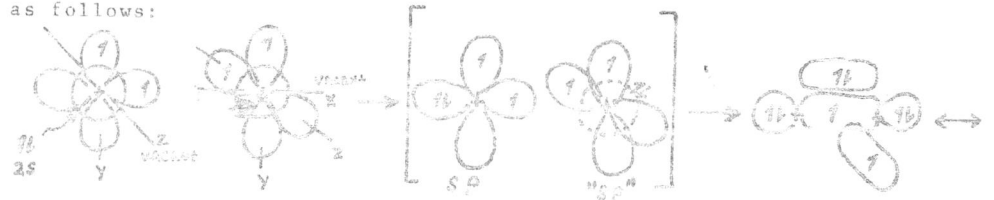

   In this configuration one atom is sp hybridized
while the other is "sp" (merged 2s and vacant p_x).
One sp orbital overlaps with the "sp" orbital
of the second atom forming a "sp", sp molecular

orbital possessing a single electron (resonance forms are possible). The y-axis and z-axis p orbitals overlap side-to-side to form π orbitals. One π orbital contains one electron, while the other possesses an electron pair. The non-bonded sp orbital has an electron pair as does the σ 2s, or σ "sp" orbital.

The bond order for this structure is two and it has two unpaired electrons (paramagnetic).

Electron-dot structure. $:C \overset{\cdot}{\div} C:$

6. N_2^+ (Diatomic Nitrogen Cation)

The N_2^+ molecular species can be "built" by combining a nitrogen ion with four valence electrons (N^+) with a neutral nitrogen atom (five electrons). The N^+ ion is isoelectronic with carbon and can be involved in bond formations similar to those encountered previously; the neutral nitrogen would have to be sp hybridized since there is an occupied x-axis p orbital. The most likely electronic configuration "ground state" for N_2^+ can be constructed as follows:

This configuration is consistent with experimental evidence for the N_2^+ species. The bond order is 2.5 and it is paramagnetic.

Electron-dot structure. $\left[:N \overset{\cdot}{\underset{\equiv}{\equiv}} N: \right]^+$

(Note. For simplicity, some p orbitals will be indicated by lines, as above.)

7. N_2 (Diatomic Nitrogen)

Diatomic nitrogen is a very stable molecule, which is also predictable from its expected electronic configuration. N_2 is the simplest system where both atoms must be sp hybridized; this is true because there is no vacant p orbital in either nitrogen atom. The electronic configuration model can be given as follows:

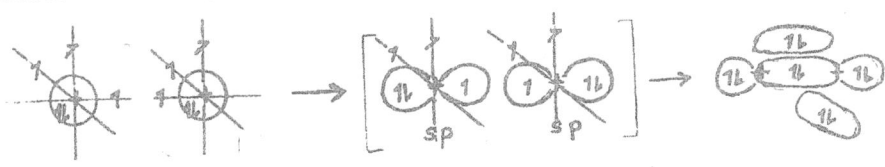

The inner sp orbitals overlap to yield a sigma sp-sp bonding (σsp) molecular orbital. The other two pair of σsp electrons are non-bonding electrons (σnb) because they counterbalance each other; also, their repulsive forces help keep bonding electrons between the two nuclei. The y-axis and z-axis p orbitals overlap to form πy and πz bonding orbitals, each containing an electron pair.

A stability zenith is reached with N_2 for diatomic molecules. The bond order is three and it is diamagnetic.

Electron-dot structure. $:N \equiv N:$

Octet Rule. As in the atomic state each atom of a molecule has an electron sphere, i.e. a volume into which electrons are pulled by the nucleus, the pulling power being related to effective nuclear charge. (Greater the effective nuclear charge, smaller the volume. For small atoms, such as those of row two elements, the electron sphere is sufficiently large (normally) for only eight (four pair) electrons. (Electron repulsive

forces keep more than eight electrons from existing in the sphere.) The electron pairs are always as far from each other as possible, which ideally means the corners of a tetrahedron.

Although each atom of N_2 has only two areas of electron density, the six bonding electrons between the nuclei still approximates the base of a tetrahedron. This is true whether these electrons are "localized" in σ and π orbitals or whether they are revolving:

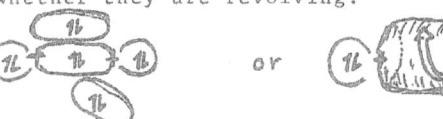 or

The N_2 molecule is stable because each atom possesses the full compliment of eight electrons and the two nuclei are held firmly "together" by six bonding electrons. Thus, we begin to understand why N_2 can be very stable while B_2 is "reactive"!

8. CO (Carbon Monoxide); NO^+ (Nitric Oxide Cation or Nitrosonium Ion)

Carbon monoxide and the nitrosonium ion are isoelectronic with nitrogen. However, they are reactive because of their polarity. The configuration can be given as follows:

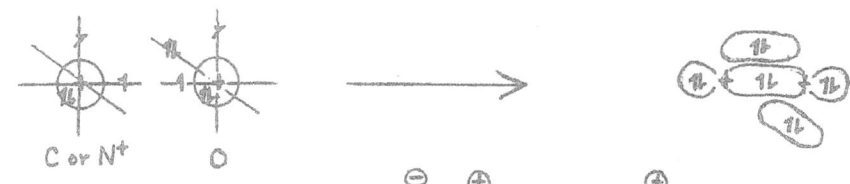

Electron-dot Structures:

[Note. Bonds between unlike atoms are unsymmetrical along x

B. Dinuclear Systems With Greater Than Ten Valence Shell Electrons

If a two-nuclear system involving second row elements has
more than ten valence shell electrons, the bond order has to be
less than three, i.e. the two atoms must share less than six
electrons. The number of electrons shared, or bond order, is related
to the total number of electrons. This relationship exists because
of the Octet Rule. As stated previously, each atomic "sphere" of
a second row element possesses no more than eight electrons in
its valence shell; the electron repulsive forces are too great
for more than eight electrons (four pairs) to exist! Therefore,
these additional electrons (number over ten) are anti-bonding
electrons in that they cause a decrease in the number of bonding
electrons relative to N_2.

9. NO (Nitric Oxide), N_2^- (Nitrogen Anion), O_2^+ (Oxygen Cation)

The NO molecule possesses eleven valence shell electrons and
can be viewed as a N_2 molecule with one additional electron (or
as the O_2 molecule with one less electron). The most recently
added (M.R.A.) electron must go into a π^* (anti-bonding)
molecular orbital region, represented as $\uparrow^*$ in A below. The M.R.A.
electron is anti-bonding because it effectively cancels one-half
of a π 'bond, i.e. it is destabilizing relative to the N_2 electronic
system.

[The M.R.A. of the NO molecule gives the oxygen nine electrons,
which violates the Octet Rule. The increased electron repulsion
created by the M.R.A. electron produces a weakened bonding system,
i.e. there is less bonding electron density in NO than in N_2!]

The configuration for NO can be given as follows:

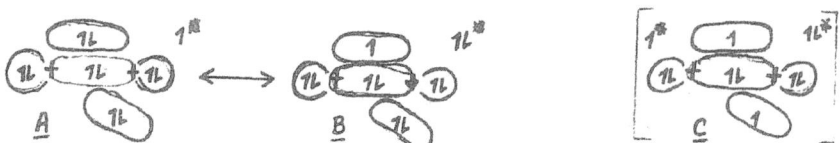

In structure $\underline{A}$ the M.R.A. is indicated as 1^{*} , which alters the
stability of a π electron pair. Structure $\underline{B}$, which is a resonance form of $\underline{A}$, shows an electron pair in the π^{*} orbital oxygen.
This form ($\underline{B}$) is likely favored since oxygen is more electronegative than nitrogen. A third form, which might exist as an
excited state species, is placed in brackets and labeled $\underline{C}$ above.
Groundstate NO is known to have a bond order of about 2.5 and
to possess only one unpaired electron; therefore, structure $\underline{B}$
above is the preferred structure.

The configuration for NO ($\underline{B}$) can also be obtained through a
linear combination of the individual atomic orbitals; the procedure
is the same as for previous systems.

Both the nitrogen and oxygen atoms are sp hybridized since
there is electron density in each p orbital. The configuration
for the NO system can be constructed as follows:

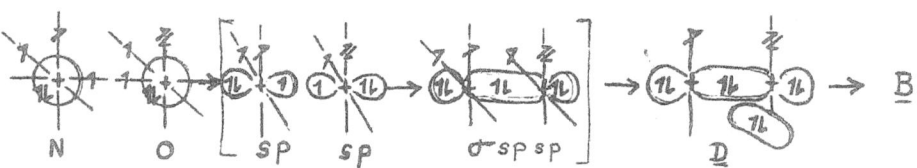

The intermediate stages are shown inside the brackets. The
sp hybrid orbitals overlap to produce a bonding σspsp molecular
orbital; also, there exists two electron pairs in the outer sp
non-bonding orbitals. (These two orbitals are σ non-bonding,

rather than a σ, σ^* combination, because of the σ spsp bonding
electrons. Only one pair of electrons may occupy the σ bonding
area!)

The intermediate structure above labeled σ spsp has an
unpaired electron in a z-axis p orbital of each atom. These
z-axis p orbitals overlap side-to-side to yield a πz bonding
molecular orbital, illustrated in structure D above.

The NO molecule could have structure D which would have a
bond order of two; the y-axis p orbitals are non-bonded, possessing
one and two electrons, respectively. However, as stated above,
experimental data indicate a bond order of about 2.5 for NO.
Apparently, the y-axis p orbitals overlap forming a πy molecular
orbital possessing a single electron; the p orbital electron pair
of oxygen is promoted (displaced) to a π^* orbital of oxygen.
Structure B, as seen above, is the preferred electronic configuration for NO. [Although the Octet Rule is violated in B for
the oxygen atom, the energy gain from πy formation evidently
"overrides" this factor.) Also, y-axis p orbital repulsion in
structure D is alleviated.]

Electron-dot structure. :N≡O:

10. O_2 (Diatomic Oxygen)

The diatomic oxygen molecule with 12 valence electrons is an
interesting study in electronic configuration. The molecule,
which is known to have a bond order of about two and two unpaired
electrons, must have four electrons in pi anti-bonding (π^*) orbitals.
The electronic configuration can be illustrated as follows:

The atoms are sp hybridized. An electron pair in a y-axis
p orbital of one atom and a pair in a z-axis p orbital of the other
atom are displaced to $\pi^\#$ orbitals. The driving force is molecular
orbital formation; each of these orbitals has a single electron,
i.e. Hund's Rule is obeyed. If the π bonding orbitals possessed
two electrons each the Octet Rule would be violated. This configuration is consistent with experimental data for O_2. The bond order
is about two and it is paramagnetic.

Electron-dot structure. $:\overset{\cdot\cdot}{O} \overset{\cdot}{\rightarrow} \overset{\cdot\cdot}{O}:$

[The $\pi^\#$ electrons can occupy any component of the two $\pi^\#$ orbitals,
which roughly surround the σnb electrons. Therefore, the $\pi^\#$
electrons have considerable freedom of movement and may carve out
a "cone" shape, with the σnb electrons "in the cone".]

11. F_2 (Diatomic Fluorine)

Fluorine with 14 valence electrons has the following electronic
configuration:

MOLECULAR ORBITAL ENERGY DIAGRAM AND TEMPLATE

Molecular Orbitals

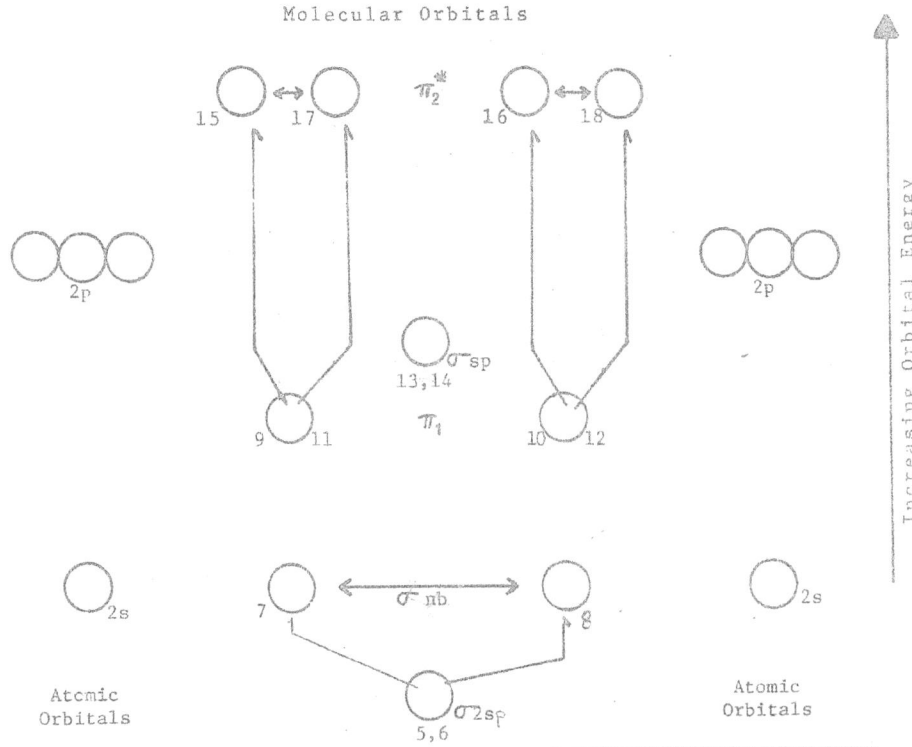

Increasing Orbital Energy

Atomic Orbitals

Atomic Orbitals

Atomic First Shells Are "Closed" in Species with Four or more Electrons (KK)

Notes.

1. The order of orbital filling is from bottom to top, i.e. increasing orbital energy. The numbers at bottom of circles correspond to most recently added (M.R.A.) electron, and the number of electrons. (The 1s orbitals, and corresponding molecular σ orbitals, are not shown since the pattern is the same as for the 2s system; the first shells are "closed" for systems with more than four electrons and are represented as KK in electronic configuration designations.)

2. The location of an electron in an orbital area that is different from a "precursor" species is indicated by a fishhook ($\rightarrow$). One electron moves from one orbital to another orbital.

3. The two σ nb (sigma non-bonding) orbital components connected by a double-headed arrow ($\leftrightarrow$) are equivalent to a σ (bonding) and σ^* (anti-bonding) combination. Either designation could be used for the four electrons in these orbitals; however, placing electrons in the σ nb orbitals circumvents the question of too many (eight) electrons in the sigma bonding area (for 13 electron systems and higher).

4. Example.
A dinuclear species with 16 total electrons (O_2, etc) can be assigned the following configuration using this diagram: $KK\ \sigma nb^4\ \pi_1^2\ \sigma sp^2\ \pi_2^{*4}$

The two most recently added (M.R.A.) electrons go into π_2^* orbitals; next an electron is promoted from each of the π_1 orbitals into the same π_2^* orbitals. These promotions are necessary if both Hund's Rule and the Octet Rule are obeyed! The bonding electrons are in the σ sp orbital and the two π_1 orbitals.

Electron-dot structure. $:\ddot{O}\dot{+}\ddot{O}:$

5. When electrons 19 and 20 add to a dinuclear species the remaining bonding electron pair (σsp) is disrupted. The atomic states (2s plus 2p orbitals) are re-established since this energy situation is now favored.

CHAPTER 3

POLYNUCLEAR ELECTRONIC CONFIGURATIONS

The Linear Combination of Atomic Orbital-Molecular Orbital
(L.C.A.O.-M.O.) method can also be extended to polynuclear systems.
The process is the same as for construction of dinuclear systems,
i.e. trinuclear systems are constructed by adding an atomic system
to dinuclear systems, tetranuclear systems are constructed by
adding an atomic system to a trinuclear system, and so forth.
THE COMBINING SPECIES (MULTINUCLEAR SYSTEM AND ATOMIC SYSTEM)
WILL HAVE THEIR CONFIGURATIONS ALTERED ONLY TO THE DEGREE NECESSARY
FOR A LOW ENERGY SYSTEM TO FORM, AND EXIST (NOTE ROOM
TEMPERATURE SPECIES IS NOT NECESSARILY THE LOWEST ENERGY SPECIES).

I.   Electronic Configuration of Some Polynuclear Systems.

   1.   NO + Cl ⟶ NOCl

The electronic configuration of NO was given in Chapter 2 (p49).
The linear combination of NO and Cl can be depicted   as follows:

A above shows the "single-electron" p orbitals of the chlorine
atom and NO interacting to produce the intermediate complex, labeled
B. (The dotted area in B outlines the p lobe that no longer possesses
electron density. (The 2s electrons of Cl are now in a σ^*sp, or
σnb, orbital.) Intermediate structure B is unstable because the
VSEPR rule is violated, i.e. the non-bonded p lobe area is void of
electron density giving the electron density around N a

"T-shape." Consequently, the sp and p orbitals of nitrogen "hybridize to sp^2, giving the correct structure (<u>C</u>) for NOCl.

sp^{2}Hybridization. This type hybridization is dictated when a second atom associates with one lobe of a p orbital of an sp hybridized speci The bonding atom distorts the p orbital: the "bonding lobe" becomes more important (longer but thinner) whereas the "backside" lobe essentially disappears, becomes devoid of electron density, i.e. the p orbital no longer exists. The result would be a "T-shape" as seen in <u>B</u> above. However, this form is structurally unstable since the V.S.E.P.R. rule is not obeyed. Consequently, the non-bonding sp (σsp$_{nb}$) electrons and the double bond electrons (σsp+πp) are "pushed" into this vacated p lobe area by the "newly bonding" electrons That is, sp^2 hybridization occurs, which can be illustrated as follows:

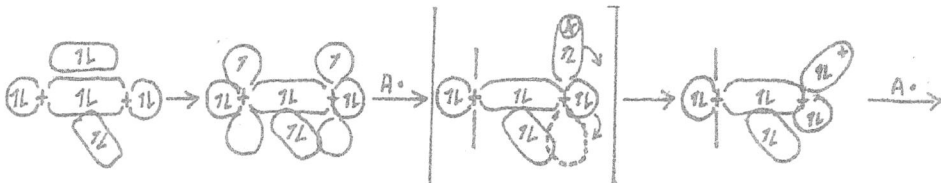

Distorted Bond Angles. Since the three areas of electron density around N in <u>C</u> are not identical, the trigonal shape is slightly distorted: the O—N—Cl bond angle is less than 120° (~117°). This is true because a non-bonded electron pair exerts greater repulsion (requires a greater volume than does a bonding electron pair), and can be visualized as follows:

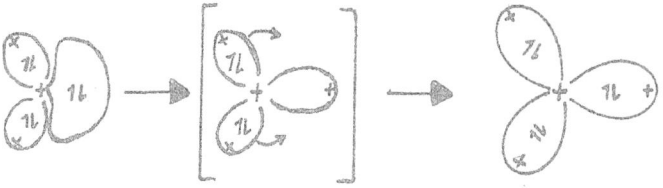

13. CH_2 + H $\longrightarrow$ CH_3

You will recall that there are two structural forms of CH_2-
linear and trigonal (p. 67). We can use each of these forms to
yield the same CH_3 structure.

A. The sp, or linear form of CH_2 can add a hydrogen atom to one of
its two non-bonded p orbitals as follows:

In the above sequence the "backside" x-axis p lobe essentially
disappears as the hydrogen atom bonds; This leaves an unbalanced
"T"-shaped structure (V.S.E.P.R. is not obeyed), shown in brackets.
The two existing hydrogens move forward into the vacated area
giving a symmetrically balanced CH_3 structure.

B. An interesting question arises when a hydrogen atom adds to an
sp^2 hybridized CH_2 species: does the hydrogen atom approach the
carbon nuclear core or does it associate with the non-bonded
electron pair?

Chemists usually treat an electron pair as a unit in their
mechanistic schemes, and it serves them well. However, it is
important to remember that the electron pair is probably closer
to the representation in A below. Since one side of a hydrogen
atom nucleus is always exposed it is likely that the hydrogen
atom associates with the electron "pair". The construction of
CH_3 from sp^2 CH_2 can be illustrated as follows:

sp^3 Hybridization. A vast array of molecules, especially organics, are either sp^2 or sp^3 hybridized. We have already looked at sp^2 hybridization; we will now look at just a few sp^3 hybridized molecules.

When the third (and last) p orbital of an atom, particularly of second row elements, becomes involved with a second atom sp^3 hybridization results. (In other words there are now four atoms associated with the central atom.) As previously, the atom associates with one lobe of the p orbital; the other lobe becomes essentially devoid of electron density, i.e. it no longer exists. The resulting electron deficient area causes the V.S.E.P.R. rule to be violated and, subsequently, a shift in position of the other electron density areas yields a tetrahedral shape, i.e. four orbitals, called sp^3 orbitals, are produced.

16. CH_3 + H $\longrightarrow$ CH_4

In the sequence above a hydrogen atom associates with one p lobe of CH_3, leaving the other lobe essentially vacant of electron density. The resulting species, which is under strain due to an inbalance of electron densities, is shown in brackets. The three planar hydrogens move downward (with electron densities) into the vacated area. The result is a symmetry balanced, tetrahedral CH_4 molecule with four equal H-C-H bonds angles of 109°28'.

This is true because the added electrons create disruptive forces
in the system, i.e. electron density becomes too crowded and a
"de-sharing" of electron density occurs (see Octed Rule p. 47).

(Appendix)   Stability Plot for Two-Nuclear Systems

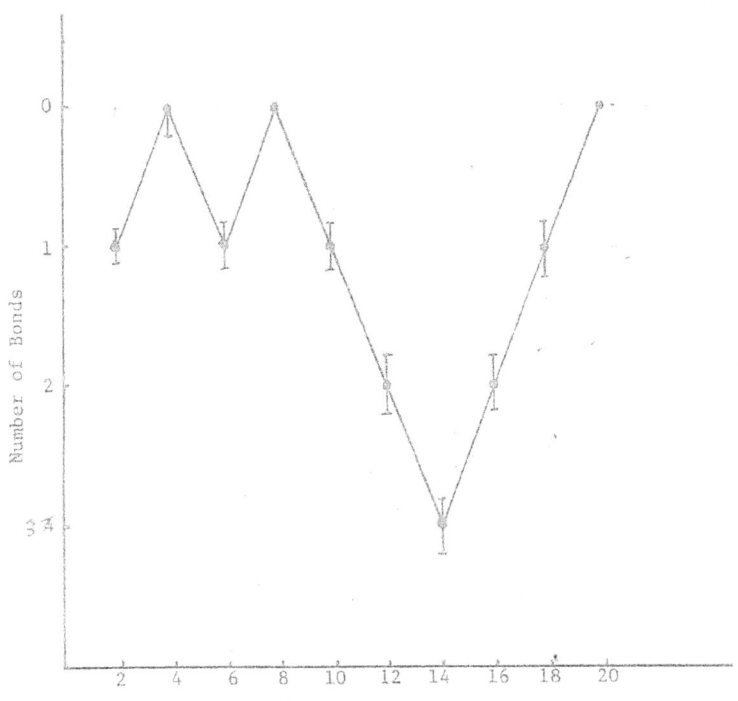

Total Number of Electrons

Total No. Electrons	Bond Order	Elements	"Constructed" Structures
1	?	H,H	
2	1	H,H	
3	½	He,H	
4	0	He,He	
5	½	Li,He	
6	1	Li,Li	
7	½	Be,Li	
8	0	Be,Be	
9	½	B,Be	
10	1	B,B	
11	1½	C,B	
12	2	C,C	
13	2½	N,C	
14	3	N,N	
15	2½	O,N	
16	2	O,O	
17	1½	F,O	
18	1	F,F	?
19	"½"	Ne,F